IGUANAS

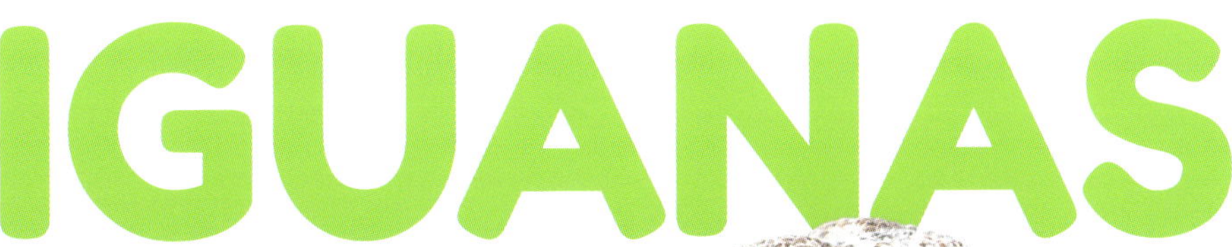

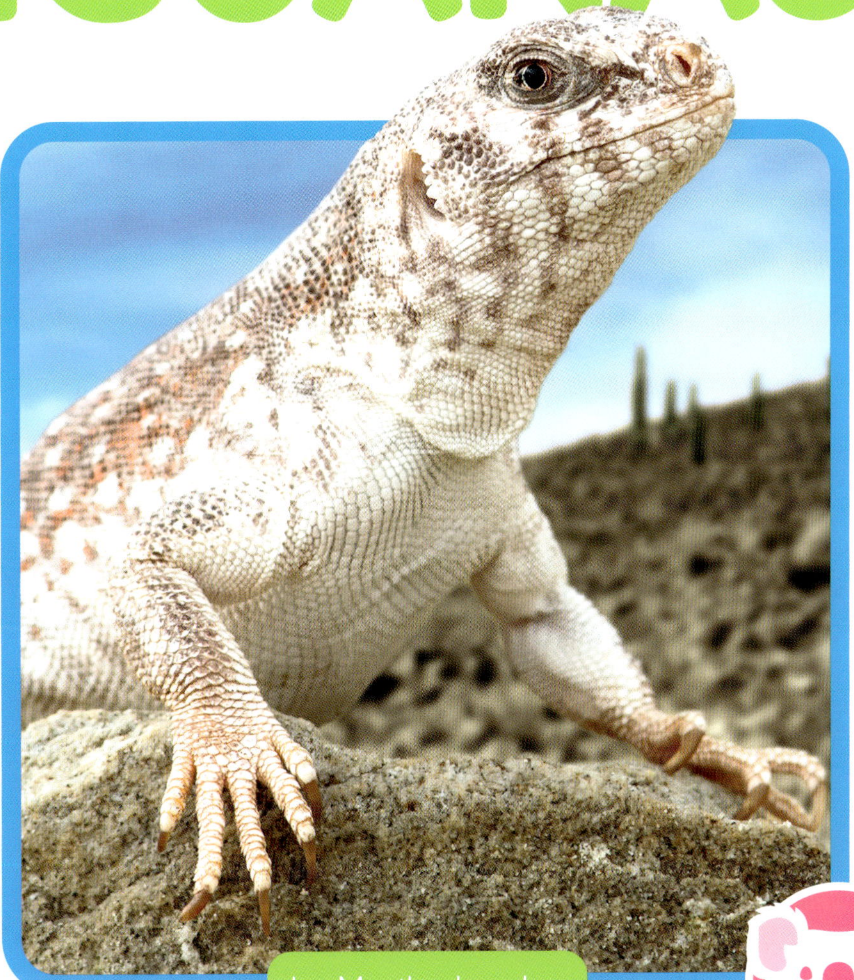

by Martha London

Cody Koala
An Imprint of Pop!
popbooksonline.com

abdobooks.com
Published by Pop!, a division of ABDO, PO Box 398166, Minneapolis, Minnesota 55439.

Printed in the United States of America, North Mankato, Minnesota.

052021
092021

THIS BOOK CONTAINS RECYCLED MATERIALS

Cover Photo: Shutterstock Images, foreground, background
Interior Photos: Shutterstock Images, 1 (foreground), 1 (background), 5, 6, 9 (top), 9 (bottom left), 9 (bottom right), 10–11, 13, 14, 17, 20; Dan Suzio/Science Source, 15; Chris Mattison/Nature Picture Library/Alamy, 18–19

Editor: Aubrey Zalewski
Series Designers: Laura Graphenteen and Colleen McLaren

Library of Congress Control Number: 2020948278
Publisher's Cataloging-in-Publication Data
Names: London, Martha, author.
Title: Iguanas / by Martha London
Description: Minneapolis, Minnesota : Pop!, 2022 | Series: Desert animals | Includes online resources and index.
Identifiers: ISBN 9781532169694 (lib. bdg.) | ISBN 9781098240622 (ebook)
Subjects: LCSH: Iguanas--Juvenile literature. | Lizards--Juvenile literature. | Lizards--Behavior--Juvenile literature. | Desert animals--Juvenile literature.
Classification: DDC 591.754--dc23

Hello! My name is

Cody Koala

Pop open this book and you'll find QR codes like this one, loaded with information, so you can learn even more!

Scan this code* and others like it while you read, or visit the website below to make this book pop.

popbooksonline.com/iguanas

*Scanning QR codes requires a web-enabled smart device with a QR code reader app and a camera.

Table of Contents

Chapter 1

Iguanas in the Sun

Desert iguanas are medium-sized lizards. They live in the southwestern United States. They also live in northern Mexico.

Watch a video here!

Iguanas are **cold-blooded**. They sit in the sun to stay warm. Desert iguanas are awake during the day. They move around in the heat.

Desert iguanas move most when it is more than 100 degrees Fahrenheit (38°C).

Chapter 2

What Do They Look Like?

Iguanas have claws on their feet. They use their claws to climb branches and rocks. Scales cover iguanas. The scales protect them from sharp rocks and sticks.

Learn more here!

Desert iguanas are brown or gray with spots. Their long tails have small dots that are reddish brown. Iguanas blend in with their **habitat**. Their colors match the rocks and sand.

tail

Iguanas' long back toes and claws help them grip branches when they climb.

Chapter 3

Running and Climbing

Birds, foxes, and snakes eat desert iguanas. Desert iguanas escape **predators** by running very fast. Sometimes they run into **burrows**.

Learn more here!

Desert iguanas stay near creosote (KREE-uh-soht) bushes. They climb these to find food.

Desert iguanas eat the leaves and flowers. They also live in burrows under the bush.

Chapter 4

Life of an Iguana

Winter is too cold for desert iguanas. They **hibernate** through the winter. They stay in **burrows** made by other animals.

Complete an activity here!

Desert iguanas **mate** in May and June. Female iguanas lay three to eight eggs.

Desert iguanas before mating

Iguanas bury the eggs. This hides them from **predators**.

Desert iguanas' bellies turn pink when it is time to mate.

Baby iguanas crawl out of burrows when they hatch. Desert iguanas do not raise their young. Baby iguanas take care of themselves. They live up to seven years.

Making Connections

Text-to-Self

Imagine you are on a hike in the desert. Would you like to see a desert iguana? Why or why not?

Text-to-Text

Have you read a book about another kind of lizard? How is that lizard like the desert iguana? How is it different?

Text-to-World

Iguanas blend into their habitats. Can you think of other animals that match their habitats?

Glossary

burrow – a hole that an animal digs in the ground for shelter.

cold-blooded – describing an animal whose body temperature is controlled by its surroundings.

habitat – the area where an animal normally lives.

hibernate – to enter a resting state during winter.

mate – to come together to have babies.

predator – an animal that hunts other animals for food.

Index

Online Resources

popbooksonline.com

Thanks for reading this Cody Koala book!

Scan this code* and others like it in this book, or visit the website below to make this book pop!

popbooksonline.com/iguanas

*Scanning QR codes requires a web-enabled smart device with a QR code reader app and a camera.